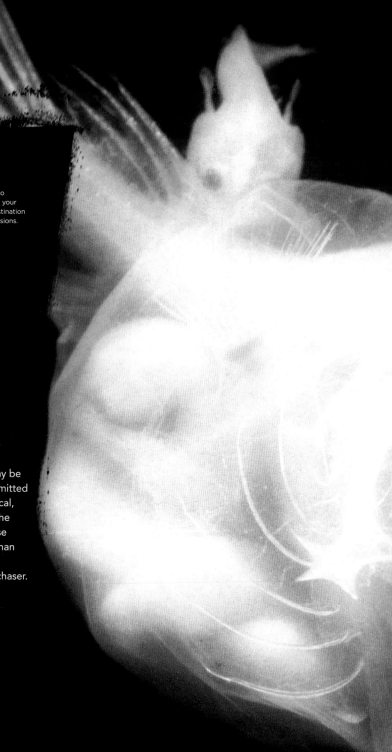

Brimming with creative inspiration, how-to projects, and useful information to enrich your everyday life, quarto.com is a favorite destination for those pursuing their interests and passions.

© 2022 Quarto Publishing Group USA Inc.

First published in 2022 by QEB Publishing,
an imprint of The Quarto Group.
100 Cummings Center, Suite 265D
Beverly, MA 01915, USA.
T (978) 282-9590
F (978) 283-2742
www.quarto.com

Author: Camilla de la Bédoyère
Editorial: Emily Pither
Design: Susi Martin
Picture Research: Sarah Bell

All rights reserved. No part of this publication may be reproduced, stored in a retrieval system, or transmitted in any form or by any means, electronic, mechanical, photocopying, recording, or otherwise, without the prior permission of the publisher, nor be otherwise circulated in any form of binding or cover other than that in which it is published and without a similar condition being imposed on the subsequent purchaser.

A CIP record for this book is available
from the Library of Congress.

ISBN 978 0 7112 6257 7

Manufactured in Guangdong, China TT032022

9 8 7 6 5 4 3 2 1

CONTENTS

CROCODILE	4
DRAGONFLY	6
BARN OWL	8
WALRUS	10
OCTOPUS	12
KOMODO DRAGON	14
VAMPIRE BAT	16
PRAYING MANTIS	18
LION	20
JUMPING SPIDER	22
SCORPION	24
STAR-NOSED MOLE	26
CHAMELEON	28
JELLYFISH	30
PIT VIPER	32
MANTIS SHRIMP	34
ALLIGATOR SNAPPING TURTLE	36
GREAT WHITE SHARK	38
SCUTIGERA	40
ANT	42
SEA ANEMONE	44
DEEP SEA ANGLERFISH	46
PICTURE CREDITS	48

CROCODILE

A crocodile has mighty jaws that can snap and grab in seconds. It uses its sharp teeth to bite down hard on its victim. The teeth lock together perfectly, so even a slippery fish can't escape.

When a baby crocodile hatches, it uses a special tooth to break the shell and escape from the egg.

A crocodile has at least 60 teeth in its long, strong snout and when they fall out new ones grow. A 70-year-old crocodile may have grown more than 2,700 teeth in its life!

FUN FACT
Crocodiles can bite... but they can't chew!

DRAGONFLY

A dragonfly may be small and dainty, but it has huge eyes and can see colors that are invisible to us! These hungry hunters lurk around ponds, where they search for other flying bugs to eat.

Dragonflies can fly fast, but they can also hover, dive, and swoop as they attack and grab their prey.

Each large eye is made up of thousands of tiny parts, which help the insect see anything moving nearby, even behind it.

FUN FACT

Dragonflies have been around for 300 million years!

BARN OWL

The heart-shaped face of a barn owl is an amazing secret weapon. It collects any tiny sound and sends it to the owl's ears. This bird has such amazing hearing it can find its victim in almost total darkness.

The fluffy edges of an owl's soft feathers muffle the sound of the bird's wings as it swoops. They use their strong claws —called talons—to catch prey.

WALRUS

This big beast's bristly whiskers help it to hunt for food on the seabed. As a walrus snuffles slowly along the seabed it uses its whiskers to feel for animals living there.

FUN FACT
Up to 700 stiff hairs grow out of a walrus's snout.

A male walrus is a huge predator that can weigh three times more than a polar bear! It can eat more than 6,000 clams a day!

When the walrus finds food it blows water out of its nose. This clears away sand and mud so the walrus can suck up the shellfish and swallow them whole!

OCTOPUS

This eight-armed animal is a wonder of the underwater world. Octopuses may have soft bodies, but they are fearless predators with plenty of brain-power.

An octopus uses its arms to swim, walk, climb, and crawl. It can also use them to grip, twist, and hold fish or shellfish.

An octopus has one big brain in its head, and eight more little brains—one in each arm!

FUN FACT

Large octopuses are strong enough to catch and kill sharks!

Each arm is lined with up to 200 strong suckers.

KOMODO DRAGON

This fearsome beast is the world's largest lizard, and it has a big appetite! Komodo dragons use a long, forked tongue to find something tasty to eat, and kill their prey with a strong bite and deadly venom.

Komodo dragons make strong venom in their mouths. They are not fussy eaters—they even like smelly, rotting meat.

FUN FACT

After one big meal, a dragon may not need to eat again for weeks.

Lizards flick their tongues to taste and smell the air.

Baby dragons climb trees to hide from adult dragons, which will hunt them if they are hungry.

VAMPIRE BAT

With its sharp fangs and black, beady eyes this vampire bat looks super scary! It uses its incredible senses to find large animals, such as cows and horses, so it can drink their blood.

In the dead of night, vampire bats sniff out their victims, and even listen for the sounds of their breathing. The bat's nose can sense heat and helps it find the best place to bite.

Sharp fangs are perfect for cutting skin, while the vampire's long tongue licks up the blood that oozes out.

FUN FACT

Each bat needs to drink about two tablespoons of blood every day.

PRAYING MANTIS

A mantis moves with lightning speed to trap its prey in its front legs. The legs are lined with dagger-like spines so they can grip tightly onto a struggling bug.

Mantises eat their prey alive, using strong jaws to crunch their meal into smaller pieces.

FUN FACT

A mantis swivels its head so it can look for danger in almost every direction.

Praying mantises are often green or brown so they can hide from their prey. When a mantis spies a victim, its front legs shoot out and snap shut around the bug.

LION

All cats use stealth, strength, fangs, and claws to catch their prey, but lions are special because they use teamwork too. They hunt in groups, circling a victim before turning on the speed in a deadly chase.

Lions use their sharp claws to catch their prey. Then they use their long fangs to take the first, deadly bite.

FUN FACT

Did you know that female lions make the best hunters?

JUMPING SPIDER

A jumping spider uses its amazing eyes to find bugs and other spiders to eat. A jumping spider has eight eyes. There are four huge eyes on the spider's face and four tiny ones on the top, or side, of its head.

FUN FACT
They can jump up to 6 inches—that's like a person jumping more than 65 feet!

Jumping spiders can see blue and orange, but not red.

Many spiders catch their food in a sticky web of silk, but jumping spiders are hungry hunters that go looking for food—and when they find it, they pounce!

SCORPION

FUN FACT Some scorpions have venom that can cause great pain, or even death, in humans.

Scorpions have eight legs because they belong to the same family as spiders. Unlike spiders, they have venom in their tails, which they inject with a sharp stinger.

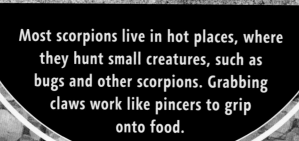

Most scorpions live in hot places, where they hunt small creatures, such as bugs and other scorpions. Grabbing claws work like pincers to grip onto food.

When a scorpion is ready to attack it raises its long, slender tail over its head and plunges the stinger into its victim.

STAR-NOSED MOLE

When this blind mole burrows in damp soil its nose is working extra hard! It has 22 soft tips that wiggle around, touching and sniffing everything.

A wiggling nose helps the mole find worms by touching up to 12 different places every second.

> Star-nosed moles are some of the world's fastest eaters, gobbling down a whole worm in the blink of an eye.

FUN FACT
This mole can smell underwater, by blowing bubbles and sniffing them back into its nose!

CHAMELEON

Chameleons are lizards with a secret weapon hidden in their mouths! When a bug passes by, a chameleon shoots out its sticky tongue, which grabs onto the bug.

The tongue flies back into the lizard's mouth like a piece of elastic. It's one of the fastest movements in the animal kingdom and the bug has no chance of escape.

Chameleons have scaly skin that can change color in a flash!

FUN FACT
A chameleon's eyes can roll around in all directions so they can spot danger, or spy bugs.

JELLYFISH

A jellyfish looks harmless as it slowly floats through the ocean but it's a soft-bodied killer! The long ribbons that dangle beneath the body are called tentacles and they are covered in tiny stingers.

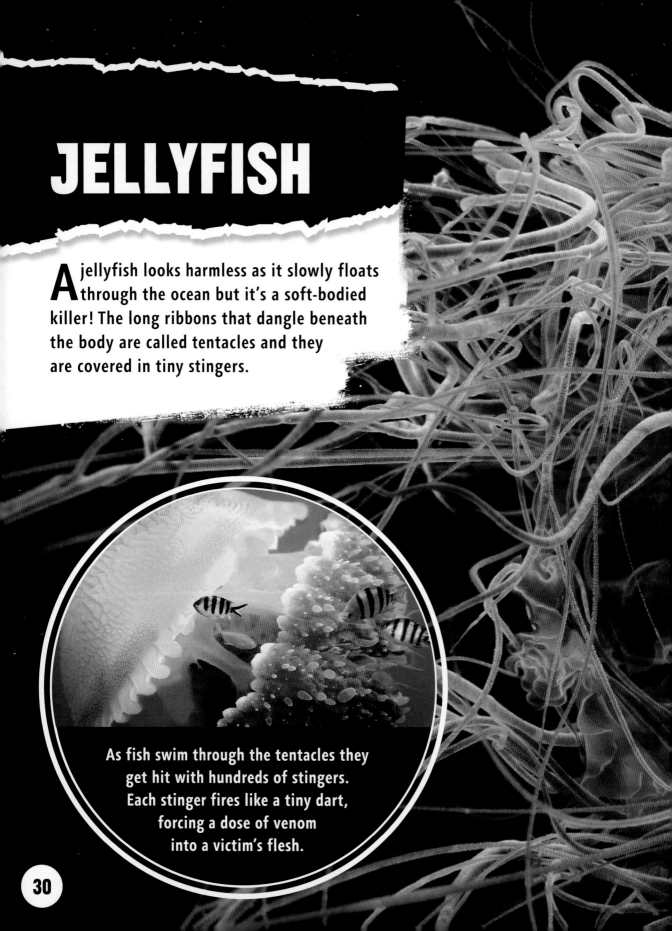

As fish swim through the tentacles they get hit with hundreds of stingers. Each stinger fires like a tiny dart, forcing a dose of venom into a victim's flesh.

The jellyfish's body is called a bell.

There are thousands of stinging cells on each tentacle. Some jellyfish venom is deadly to humans.

FUN FACT

This ocean animal does not have any bones or a brain.

PIT VIPER

Look at the dark pit between this snake's eye and its nostril. It holds the secret to this reptile's great success as a hunter—but this pit viper has other lethal weapons too!

A pit viper senses the heat given off by warm-blooded animals. It senses prey even in the dark and strikes quickly, using its long fangs to bite and inject venom.

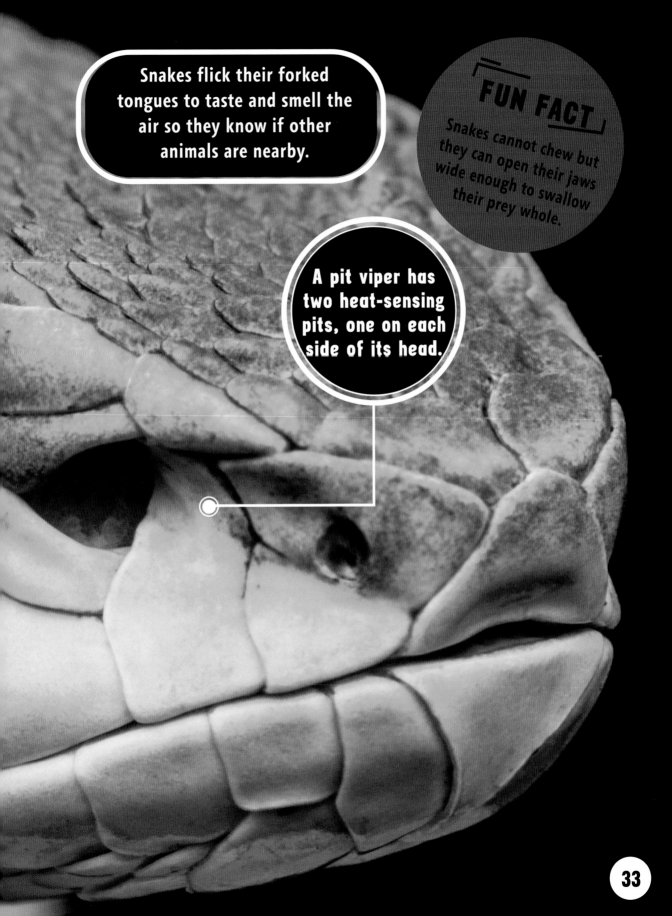

Snakes flick their forked tongues to taste and smell the air so they know if other animals are nearby.

FUN FACT

Snakes cannot chew but they can open their jaws wide enough to swallow their prey whole.

A pit viper has two heat-sensing pits, one on each side of its head.

MANTIS SHRIMP

With great eyesight and a powerful punch, this little shrimp is a deadly killer. It hides on the seabed, where it looks out for fish or shellfish swimming past.

A mantis shrimp can see more colors than any other animal and its tough skin protects its body, like a suit of armor.

FUN FACT
A mantis shrimp punches with the speed and force of a speeding bullet!

ALLIGATOR SNAPPING TURTLE

In the dark, still waters of a swamp, a turtle sits and waits. With a deadly bite and a worm-like tongue, this alligator snapping turtle is a cold-blooded killer that tricks its victims to come close.

Alligator snappers have a spiky shell, a hooked beak. and a thick, scaled tail. They spend most of their lives in water.

FUN FACT
These turtles hold their breath underwater for up to 50 minutes at a time.

A piece of bright red flesh on the turtle's tongue wiggles like a tasty worm. A tempted fish swims right up to the reptile, which suddenly snaps its strong jaws shut.

GREAT WHITE SHARK

This may not be the world's largest shark but it is packed with power, deadly weapons, super senses and a big appetite. They look scary, but great whites hunt fish and seals, not people!

A shark has several rows of teeth so it always has spares ready when old teeth fall out.

A mighty jaw helps this shark bite with a deadly strength.

FUN FACT

Sharks can feel, see, hear, smell, and taste other animals in the water.

SCUTIGERA

Bugs need to watch out for mini-monsters with lots of legs and deadly venom! Scutigeras are speedy centipedes that live in homes and forests, where they hunt at night.

FUN FACT

This centipede has 15 pairs of legs but some centipedes have hundreds!

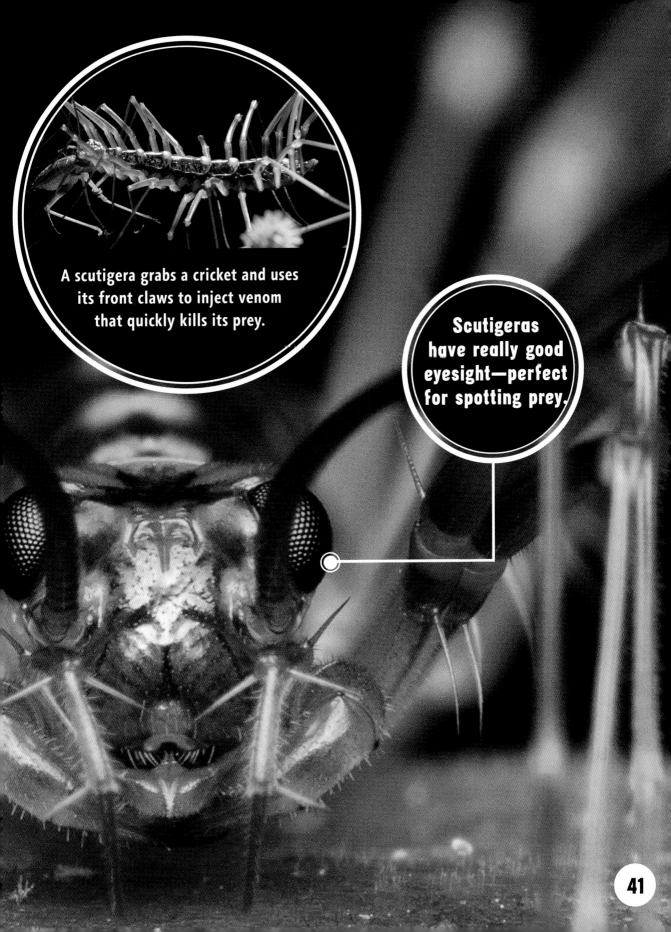

A scutigera grabs a cricket and uses its front claws to inject venom that quickly kills its prey.

Scutigeras have really good eyesight—perfect for spotting prey.

ANT

An ant can use its mighty jaws like hands to grab a victim before chewing it up into smaller pieces. Some ants also have painful stings and these little insects often hunt as a team.

An ant's chewing jaws are called mandibles.

Ants use their antennae to touch, taste, and smell. This helps them find food and track their prey.

FUN FACT

Did you know that ants carry prey heavier than themselves?

SEA ANEMONE

Colorful sea anemones look like flowers growing undersea, but they are animals with tentacles. When a fish swims past, the tentacles sting it and push it into the anemone's mouth.

FUN FACT
The world's largest sea anemones grow to around 40 inches wide.

Anemones have a clever survival technique. Bits of an anemone can break off and grow into new animals!

The tentacles grow around the anemone's mouth, which is at the top of a tube-shaped body.

DEEP SEA ANGLERFISH

Light is made here, at the end of a "fishing rod."

There is no sunlight in the deep, dark ocean so predators use some special tricks to find food. This anglerfish makes its own light, which draws other fish close to its huge mouth.

Fish are lured toward the light and the anglerfish then opens its jaws wide and sucks in its prey.

FUN FACT

Anglerfish use their fins to walk on the seabed.

47

PICTURE CREDITS

AgeFotostock
Mint Frans Lanting 8-9, Thorsten Negro/imageBROKER 9b

Alamy
Avalon/Bruce Coleman Inc 36b, BIOSPHOTO 32b,
Brandon Cole Marine Photography 38-39,
Design Pics Inc 36-37, gary tack 4-5, 6-7, Gillian Pullinger 28b,
imageBROKER 34b, Mohd Zaidi Razak 19b, Sean Chinn @
greatwhitesean 34-35, Solvin Zankl 46-47, WaterFrame 14-15

Getty Images
Dirk94025 20b, Paul Souders 10-11, 28-29, Peter K Burian 8b, seen by
tobfl 12-13, shikheigoh 32-33, Steven Trainoff Ph.D. 44-45

Nature PL
Brandon Cole 45b, Mark Moffett 41t, Michael & Patricia Fogden 16-17,
Mike Parry 38b

National Geographic
National Geographic Image Collection: 16b

Shutterstock
Amelie Koch 10b, amirhamzaa 23b, Anan Kaewkhammul 8-9, Arunee
Rodloy 4t, Cornel Constantin 22-23, DWI YULIANTO 42-43, EcoPrint
24-25, Jay Bo 20-21, kesipun 6b, Kondratuk Aleksei front cover, 12b,
Kurit afshen 50, Neil Bromhall 2-3, Paul Looyen 18-19, Ron Eldie 40-41,
Sergey Uryadnikov 14b, Vlad61 30b, Yomka 42b

SPL
Alexander Semenov 30-31, Dante Fenolio 46b, Ken Catania/Visuals
Unlimited Inc 26-27, 26b, Phil Degginger/Science Source 24b